THE TOP SECRET LIFE OF PLANTS

PLANTS THAT GROW WITHOUT SOIL

BY JANEY LEVY

Gareth Stevens
PUBLISHING

Please visit our website, www.garethstevens.com. For a free color catalog of all our high-quality books, call toll free 1-800-542-2595 or fax 1-877-542-2596.

Cataloging-in-Publication Data
Names: Levy, Janey.
Title: Plants that grow without soil / Janey Levy.
Description: New York : Gareth Stevens Publishing, 2020. | Series: The top secret life of plants | Includes glossary and index.
Identifiers: ISBN 9781538233955 (pbk.) | ISBN 9781538233979 (library bound) | ISBN 9781538233962 (6pack)
Subjects: LCSH: Epiphytes–Juvenile literature. | Plants–Juvenile literature. | Plants–Adaptation–Juvenile literature.
Classification: LCC QK922.L48 2020 | DDC 581.6′3–dc23

First Edition

Published in 2020 by
Gareth Stevens Publishing
111 East 14th Street, Suite 349
New York, NY 10003

Designer: Sarah Liddell
Editor: Abby Badach Doyle

Photo credits: Cover, p. 1 Bildagentur Zoonar GmbH/Shutterstock.com; glass dome shape used throughout bombybamby/Shutterstock.com; leaves used throughout janniwet/Shutterstock.com; background texture used throughout MInerva Studio/Shutterstock.com; p. 5 Wolfgang Kaehler/Contributor/LightRocket/Getty Images; p. 7 Matt Tilghman/Shutterstock.com; p. 9 SaveJungle/Shutterrstock.com; p. 11 Tim Graham/Contributor/Getty Images News/Getty Images; p. 13 Zulashai/Shutterstock.com; p. 15 Casa nayafana/Shutterstock.com; p. 17 Nick Pecker/Shutterstock.com; p. 19 Andreas Ruhz/Shutterstock.com; p. 21 (orchid) Marian Lazaro Martin/Shutterstock.com; p. 21 (bromeliad) Kristyna Henkeova/Shutterstock.com; p. 21 (Spanish moss) Stefan Holm/Shutterstock.com; p. 21 pangcom/Shutterstock.com; p. 21 (rain forest cactus) Cheng Wei/Shutterstock.com.

Printed in the United States of America

CPSIA compliance information: Batch #CS19GS: For further information contact Gareth Stevens, New York, New York at 1-800-542-2595.

CONTENTS

Words in the glossary appear in **bold** type
the first time they are used in the text.

MEET THE
AIR PLANTS

Plants grow in the ground, right? Their roots hold them there, taking up water and **nutrients** from the soil to keep the plant growing and healthy. But do you want to know a secret? Some plants have found a way to free themselves from a life bound to the soil.

These plants are called air plants, or epiphytes (EHP-uh-fyts). They can grow high on trees, telephone poles, or even telephone lines. They're found mostly in tropical areas.

CLASSIFIED!

EPIPHYTES HAVE BEEN AROUND FOR A LONG TIME. SOME FIRST APPEARED MORE THAN 300 MILLION YEARS AGO!

A SINGLE TREE CAN BE THE **HOST** TO MANY AIR PLANTS.

LIVING THE HIGH LIFE

Why do epiphytes choose to live high on trees and other places? That seems like a strange thing to do. Since they don't grow in soil, it seems as if they would have trouble getting the water and nutrients they need.

However, epiphytes are small plants that grow in crowded forests. With so many trees around, very little sunlight reaches the ground. If the epiphytes grew there, they wouldn't get the sunlight they need for **photosynthesis** (foh-toh-SIHN-thuh-suhs). But high on tall trees, epiphytes can grow where the sun shines.

CLASSIFIED!
THE WORD *EPIPHYTE* COMES FROM GREEK WORDS THAT MEAN "ON TOP" AND "PLANT."

THE SUN SHINES BRIGHTLY AT THE TOP OF A RAIN FOREST... BUT ON THE GROUND, OR FLOOR, IT IS VERY DARK.

RAIN FORESTS AND CLOUD FORESTS

The kinds of forests epiphytes grow in are among the secrets to their success. So it's important to know something about them. Tropical rain forests are very warm. And, as their name suggests, they get lots of rain—more than 80 inches (203 cm) per year.

Cloud forests also are in tropical areas, but they're high in mountains. This makes them cooler than rain forests. Their trees are shorter. And they're often wrapped in cloud-like fog or mist, which gives them their name.

CLASSIFIED!

CLOUD FORESTS GET LESS SUNLIGHT THAN RAIN FORESTS BECAUSE THEY'RE WRAPPED IN FOG. HOWEVER, THE FOG GIVES CLOUD FORESTS MUCH HIGHER YEARLY **PRECIPITATION**—UP TO 385 INCHES (978 CM)!

THE RAIN FOREST IS MADE UP OF MANY LAYERS.
THE CANOPY IS WHERE MOST OF THE PLANTS
AND ANIMALS LIVE.

RAIN FOREST LAYERS

EMERGENT
LAYER

CANOPY
LAYER

UNDERSTORY
LAYER

FOREST
FLOOR

9

NOT PARASITES

Epiphytes may grow on hosts, but they don't harm them like a parasite does. Parasites are living things that take nutrients from their host and don't give it anything in return. Some can make the host so weak that it dies.

Epiphytes don't take anything from their host. They simply use their host for support, to hold them up and carry them. So how, then, do epiphytes get the water and nutrients they need? They have some pretty amazing secrets, as you'll see!

CLASSIFIED!

THE CONNECTION BETWEEN AN EPIPHYTE AND ITS HOST IS CALLED COMMENSALISM (KUH-*MEHN*-SUH-LIH-ZUHM). THE EPIPHYTE GETS A BENEFIT, BUT THE HOST IS NEITHER HELPED NOR HARMED.

MISTLETOE IS A PARASITE PLANT THAT GROWS ON TREES. HERE, MISTLETOE HAS CAUSED GREAT HARM TO THIS TREE, WHICH HAS FEW LEAVES.

THE SECRET TO GETTING WATER

Epiphytes get the water they need from rain, or sometimes, even from the air! Or, to be more exact, they get water from the water vapor in the air. Most epiphytes take up water through their roots, although some can take it up through their leaves.

Some epiphytes are able to store water for dry times. They store it in thick stems and leaves. Some epiphytes have structures, or parts, to help them keep from losing water.

CLASSIFIED!

EPIPHYTES ARE SKILLED AT COLLECTING EVERY BIT OF WATER THEY CAN. THEY CAN COLLECT IT FROM THE **DAMPNESS** ON THE SURFACE OF THEIR HOST.

THESE ARE THE ROOTS OF AN EPIPHYTIC FLOW
CALLED AN ORCHID (OHR-KIHD).

1

THE SECRET TO FINDING
NUTRIENTS

It seems like finding nutrients would be trickier than finding water, but epiphytes have mastered that, as well. **Debris** made up of fallen leaves and animal waste gathers on the host and supplies nutrients for the epiphytes. Some epiphytes actually have a basket shape for catching debris as it falls.

Epiphytes such as bromeliads (broh-MEE-lee-adz) hold pools of water that are home to many small creatures, such as frogs and tadpoles. Their waste—and their rotting bodies after they die—provides nutrients for the plants.

CLASSIFIED!

YOU MIGHT BE SURPRISED TO LEARN THAT EVEN RAIN PROVIDES NUTRIENTS! IT CAN SUPPLY EPIPHYTES WITH HIGH LEVELS OF IMPORTANT **MINERALS**.

EPIPHYTES KNOW TO GROW AT A FORK IN THE TREE. THE FORK OFFERS A GOOD PLACE FOR DEBRIS TO GATHER.

INSIDE
THE ORCHID

Orchids are beautiful flowers that are popular with many plant lovers. More than 20,000 species, or kinds, exist, and almost three-quarters of those are epiphytes.

Orchids have an extra secret to help them take in and hold on to water. Their roots have a sponge-like, gray coating called the velamen (vuh-LAY-muhn). When it rains, the velamen takes up water and holds onto it. When the velamen dries out, it helps keep the inner root from losing water.

CLASSIFIED!

ORCHIDS MAKE NEW PLANTS BY PRODUCING HUNDREDS OF THOUSANDS OF TINY SEEDS. THE SEEDS ARE COVERED WITH A BALLOON-LIKE COAT, AND WIND SPREADS THEM FAR FROM THE PARENT PLANT.

16

MOST ORCHIDS MAKE FOOD THROUGH PHOTOSYNTHESIS. SOME GET NUTRIENTS FROM DEBRIS, AND SOME GET HELP FROM A **FUNGUS** THAT LIVES INSIDE THEIR ROOTS.

17

BRIGHT AND BOLD
BROMELIADS

Like orchids, bromeliads are popular with many plant lovers. They often have stiff leaves and brightly colored flowers. More than 2,000 species exist, although not all are epiphytes. Bromeliads can grow on other plants, sand, rocks, or cliffs.

Bromeliads have a couple of secrets for getting water. Tank bromeliads have **overlapping** leaves that can hold 1 quart (1 L) or more of water. The bromeliad called Spanish moss hangs from tree branches and has special hairs that take in water.

CLASSIFIED!

SPANISH MOSS ISN'T ACTUALLY SPANISH...OR A MOSS! LEGEND HAS IT THAT THIS NORTH AMERICAN BROMELIAD GOT ITS NAME BECAUSE ITS HAIR-LIKE APPEARANCE REMINDED EARLY FRENCH SETTLERS OF SPANISH EXPLORERS' BEARDS.

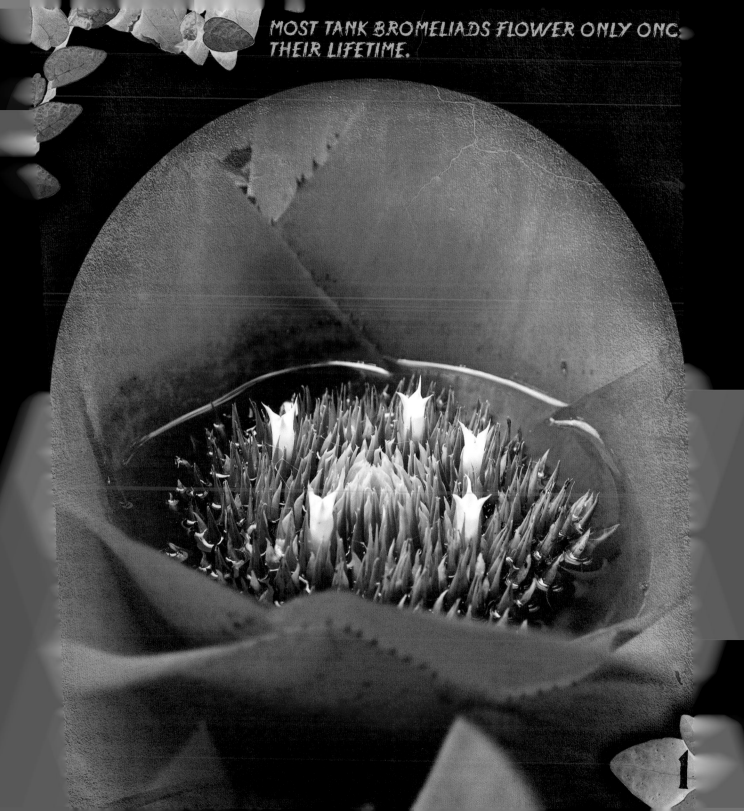

WONDERS OF ADAPTATION

When it comes to **adaptation**, epiphytes are experts. Forced to adapt by conditions in rain forests and cloud forests, they left the ground for the treetops. There, they had sunlight. But they had to find water and nutrients without soil to draw them from. However, there was rain as well as plant debris and animal waste to help provide nutrients.

Epiphytes solved their problems of finding water and nutrients with wonderful and creative adaptations. And along the way, they became beautiful and amazing plants!

EPIPHYTES

ORCHIDS

BROMELIADS

SPANISH MOSS

RAIN FOREST
CACTUS

BIRD'S-NEST
FERN

21

GLOSSARY

adaptation: a change in a type of plant or animal that makes it better able to live in its surroundings

dampness: the state of being slightly wet

debris: things that are lying where they fell

fungus: a living thing that is somewhat like a plant, but doesn't make its own food, have leaves, or have a green color. Fungi include molds and mushrooms.

host: the plant or animal on or in which an epiphyte or parasite lives

mineral: matter important in small amounts for the health of plants and animals

nutrient: something a living thing needs to grow and stay alive

overlap: to partly cover the same area

photosynthesis: how a plant makes food using water, nutrients, sunlight, and a gas called carbon dioxide from the air

precipitation: rain, snow, sleet, or hail

tropical: having to do with the warm parts of Earth near the equator

FOR MORE INFORMATION

BOOKS

Calver, Paul, and Toby Reynolds. *Rainforests*. London, UK: Franklin Watts, 2015.

Howell, Izzi. *Rainforest*. London, UK: Wayland, 2016.

McGinlay, Richard. *My First Encyclopedia of the Rainforest: A Great Big Book of Amazing Animals and Plants*. London, UK: Armadillo, 2018.

WEBSITES

12 Incredible Orchid Facts That No One Ever Told You Before
plainviewpure.com/12-incredible-orchid-facts-no-one-ever-told/
Discover some amazing and surprising facts about orchids on this website.

Epiphytes!
www.canopywatchinternational.org/2016/03/epiphytes/
Read more about air plants and see some great pictures here.

Tropical Rain Forest Facts: Bromeliad Facts
www.tropical-rainforest-facts.com/Tropical-Rainforest-Plant-Facts/Bromeliad-Facts.shtml
Learn more about the common air plants called bromeliads on this site.

INDEX